MY TEXAS TOASTMASTERS

BOOK ON TECHNOLOGY AND

COMMUNICATION

By Carla Conti Sanzone
Copyright 2015

DEDICATION PAGE

This little book on technology and communication in the modern world of the 2010's and onward is dedicated to all those in the world who utilize technology. This little book is also dedicated to anyone in the USA and world who is presently suffering from any ailment, disease, hardship, etc. May a cure, hope and help be afforded them.

CHAPTER ONE

Is there an art for staying focused in today's distracting world? That is my question to you. Continuous partial attention - a new term. This term was a term coined in the 1980's and 1990's which basically means ' the modern predicament of being constantly attuned to everything without fully concentrating on anything.

The person who 'coined' this term states that 'continuous partial attention' is neither good nor bad. She states that 'we need different attention strategies in different contexts.' The way you use your attention when you're writing a story may vary from the way you use your attention when you're driving a car, serving a meal, making love, or riding a bicycle. The important thing for us as humans is to have the capacity to 'tap the attention strategy' that will best serve us in any given moment.

From the time we're born, we're learning and modeling a variety of attention and communication strategies. For example, one parent might put one toy after another in front of the baby until the baby stops crying. Another parent might work with the baby to demonstrate a new way to play with the same toy.

That is how we are wired. The adult sticks his tongue out. The baby follows suit by sticking 'their tongue out.' The child 'mirrors' the adult.

FACT: Children learn the virtue, you may call it a virtue or quality - of 'empathy' through eye contact and gaze. If children are learning empathy through eye contact, and our eye contact is with devices, the children will miss out on empathy.

FACT: In essence, what we are doing presently is 'modeling a primary relationship with screens' and a lack of eye contact with people because of all the computers, laptops, television viewing. It ultimately can 'feed the development' of a kind of 'sociopathy and psychopathy.'

The generation that has been 'tethered' to devices serves as a 'cautionary' example to the next generation of children - those being born now - 2015 and beyond.

FACT: When we learn how to play a sport or an instrument; how to dance or sing; or even how to 'fly a plane', we learn how to 'breathe' and 'how to sit or stand' in a way that supports a 'state of relaxed presence.'

These are very different strategies and they set up a very different way of relating to the world for these children. Adults model attention and communication strategies and children imitate. In some cases, through sports or crafts or performing arts children are taught attention strategies. Some of these might be managing the breath and emotions - bringing one's body and mind to the same level. Self-directed play allows both children and adults to develop a powerful attention present called 'relaxed presence.'

Let's talk about reading or building things. When you did or do these things, nobody was giving you an assignment, nobody was telling you what to do - there wasn't any stress around it. You did these things for your own pleasure and joy. As you played, you developed a capacity for attention and for a type of curiosity and experimentation that can happen when you play. You were literally 'in the moment' and the 'moment' was unfolding in a natural way. You were, in essence, in a state of 'relaxed presence' as you explored your world.

When people talk about 'attention problems' in modern society, they usually mean the 'distractive' potential of smartphones and so on.

We learn by 'imitation' from the very start.

FACT: Mind and body in the 'same place' at the same time is 'relaxed presence.'

CHAPTER TWO
A Brief History of the Beginnings of Communication

Throughout history, developments in technology and communications have gone hand-in-hand, and the latest technological developments such as the Internet and mobile devices have resulted in the advancement of the science of 'communication' to a 'new level.'

The process of human communication has evolved over the years, with many path-breaking inventions and discoveries 'heralding' revolutions or a lift from from 'one level to another.'

FACT: The invention of pictographs or the first 'written' communication in the ancient world gave us 'written communication.' These 'writings' were on 'stone' and remained 'immobile.'

FACT: The invention of 'paper, papyrus, and wax', culminating in the 'invention of the printing press' in the 15th century made possible transfer of documents from one place to another allowing for 'uniformity of languages over long distances.'

FACT: The latest revolution is the 'widespread application of electronic technology' such as electronic waves and signals to communication, manifesting in the electronic creation and transfer of 'documents' over the World Web Web.

FACT: Speed and Costs -
The most significant impact of technology on communication is the spread of the Internet and the possibility of sending emails and chatting. In the 'pre-information technology days', a document often required re-typing on the 'typewriter' before the final version. Sending the letter across a distance to someone else required a 'visit to the post office' and a 'postage stamp.' Faster methods such as telegrams had severe limitations in text and were quite costly. Computers and the Internet have 'made the process of creating and editing documents' and applying features such as 'spell check' and 'grammar check' automatically easy and natural. E-mailing documents to any 'part of the globe' became very easy and done 'within seconds,' therefore 'making letter writing and sending out ordinary letters' somewhat OBSOLETE.

QUALITY:
The huge amount of knowledge accessible by the 'click of the mouse' has helped improve the quality of communication. Examples might be: Translating a text from an unfamiliar language to a familiar

and seeking out the 'meaning of an unknown word' and getting followup information on an 'unfamiliar concept' are all possible 'thanks to the internet.'

GOOD POINTS ABOUT TECHNOLOGY CONTINUED:

Technology 'allows' easy storage and retrieval' of communication when needed, especially 'verbal communication' the storage of which was difficult before. It now becomes 'easier to rewind and clear misconceptions' rather than make 'assumptions' or contacting the person again to 'clear doubt.'

The invention of 'new gadgets' such as mobile phones made communication easier by allowing people to communicate from 'anywhere.'

CONS ABOUT TECHNOLOGY:

The possibility of high quality communication from anywhere in the world to anywhere else at low costs ' has 'led to a marked decline in face-to-face communications' and to an 'increased reliance on verbal and written communication' over electronic mediums.

The 'small' keyboards on mobile phones and other hand-held devices that make typing difficult has 'resulted' in a radical 'shortening of words and increasing use of 'symbols and shortcuts' with 'little or no adherence to traditional grammatical rules.' Communication has 'become concise and short', and the adage 'brevity is the soul of wit' finds widespread implementation, though unintentionally.

SUGGESTION: If someone uses the internet too long or their I-phone or other technological device at work OR at home too long or longer than the other person, than the person who has used technology the most that day 'will treat all to lunch or make something foodwise.'

CHAPTER THREE

QUESTION: Is the Internet making you stupider or smarter?

TOPIC - CONCENTRATION:

Someone writes, 'my concentration' often starts to drift after two or three pages when reading a book. ' I get fidgety, lose the thread, begin looking for something else to do. I feel as if I'm always dragging my wayward brain back to the text that I am reading. The deep reading that used to come naturally has become a struggle.' Does this sound familiar to any of you?

A person by the name of Clive Thompson has written, ' the perfect recall of silicon memory can be an enormous boon to thinking, but that 'boon' comes at a price.'

A media theorist name of Marshall McLuhan pointed out in the 1960's that 'media are not just passive channels of information. They 'supply' the stuff of thought, but they also 'shape' the process of thought. And what the internet seems to be doing is 'chipping away the capacity for concentration and contemplation.' One's mind now sometimes expects to take in information the way the Net distributes it: in a swiftly moving stream of particles.'

POSSIBLE FACT: Because of the fast speed of the internet and that one has to read very quickly and fast, the process of mentally 'absorbing' this material may be too much for some people. 'People will now 'skim' reading materials such as long book passages, etc., instead of reading them fully with their eyes and interpreting the passages with their mind.'

One psychologist stated it this way, "We are not only what we read but we are how we read."
This same psychologist worries that the 'style of reading promoted by the Internet, a style that puts efficiency and immediacy above all else, may be weakening our capacity for the kind of 'deep reading' that emerged when an earlier technology, the printing press, made long and complex works of prose commonplace. When we read 'online', we tend to become 'mere decoders of information.'

FACT: Reading is NOT an instinctive skill for human beings. It is not etched into our 'genes' the way 'speech' is. We have to 'teach our minds' how to translate the 'symbolic characters we see into the language we understand.' And the media or other technologies we use in learning and

and practicing the craft of reading, play an important part in 'shaping the neural circuits inside our brains.' Experiments demonstrate that readers of ideograms, like the Chinese, develop a 'mental circuitry for reading that is very different from the circuitry found in those of us whose written language employs an alphabet.' The human brain is very complex. There are 'cognitive functions going on - memory - and also in the brain going on is 'interpretation of visual and auditory stimuli constantly.'

FACT: Isn't 'writing down with pen and paper our feelings, quotes, etc. information gathered every day, a better way to communicate at times?' 'Our writing equipment takes part in the forming of our thoughts' stated Friedrich Nietzsche in 1882. So.... isn't using a typewriter or a pen and paper QUITE different than typing on a computer keyboard? Which do you think is more effective and better?

British mathematician Alan Turing, a genius, stated, 'The Internet, an immeasurably powerful computing system, is subsuming most of our other intellectual technologies. It is becoming 'our map, our clock, our printing press, our typewriter, our calculator, our telephone, our radio and our T.V.'

Is 'scattering' our attention and diffusing our concentration every few minutes for a 'new e-mail reading, etc.' GOOD FOR US?

A gentleman in the 1800's, became part of the Industrial Revolution. Mr. Taylor used a 'stopwatch' in a steel plant in Philadelphia, and began a series of experiments aimed at improving the plant's efficiency of its machinists. Taylor timed the metalworking movements of the machinists. He 'broke down every little job' into a sequence of small discrete steps. Taylor kind of developed an 'algorithm' - for how each worker should be working. The employees complained that they felt like 'automatons' - machines - instead of human beings - but the concluding result of the experiment was that 'the factory's productivity' increased.

QUESTION: Should the blood, sweat and tears of a human being's work and toil be substituted in the future and now by a 'computer system.' Is this morally ethical?

CONS: The more computer companies and the internet in general discourage 'leisurely reading ' and 'slow concentrated thought', it is better for their economic interest to drive the average person to distraction.

FACT: The kind of 'deep reading' that a 'sequence of printed pages 'promotes is valuable not just for the KNOWLEDGE we acquire from the author's words BUT for the intellectual vibrations 'these words' set off within our minds. In the 'quiet spaces' opened up by the sustained, undistracted reading of a book, or by an act of meditation or contemplation, we make 'our own associations', draw our own 'inferences' foster and develop our own 'ideas', etc. Deep reading is indistinguishable from 'deep thinking.'
If we 'lose those quiet spaces' or fill them up with 'content', we will 'sacrifice' something important not only within ourselves BUT in our culture.

QUESTION: Should we go about life in a kind of 'robotic' existence and 'robotic' efficiency? OR......
should we do half and half or 60/40 and rely on deep reading and deep intellectual thinking with our brains and the 40% with technology and computers.
The future is up to you and only you can make that decision.

EPILOGUE

SOME QUOTES ON TECHNOLOGY AND COMMUNICATION

1. "The most technologically efficient machine that man has ever invented is the 'book.' - Northrop Frye - Canadian literary talent and theorist

2. "It has become appallingly obvious that our technology has exceeded our humanity." - Albert Einstein - Genius - Physicist

3. "One machine can do the work of 50 ordinary men. No machine can do the work of one extraordinary man." - Elbert Hubbard - American writer

4. "Technology.... the knack of so arranging the world that we don't have to experience it." - Max Frisch - Swiss playwright and novelist

5. "The great myth of our times is that technology is 'communication." - Libby Larsen - American woman music composer

6. "So much technology, so little talent." - Vernor Vingh - Prof. of Math;SciFi writer

7. "The real danger is not that computers will begin to think like men, but that men will begin to think like computers." -Sydney Harris - American journalist

8. "Technology is a useful servant BUT a dangerous master." - Christian Lange - Norwegian historian; teacher and political scientist

9. "Science and technology revolutionized our lives, but memory, tradition and myth FRAME our responses." - Arthur M. Schlesinger - American historian and intellectual

10. "Ethics change with technologies." - Larry Niven - American Science Fiction writer

11. "The real problem is not whether machines think - but whether men do." - B. F. Skinner - writer

12. "Once a new technology rolls over you, if you are not part of the steamroller, you are 'part of the road.' - Stewart Brand - American writer and author of The Whole Earth Catalog

13. THE BEST QUOTE OF THEM ALL - "The human spirit must PREVAIL over technology." - Albert Einstein

REFERENCES

1. "The Art of Staying Focused in a Distracting World" - Article from The Atlantic Monthly Magazine - 6/2013 - Author - James Fallows

2. "Is Google Making Us Stupid?" - Article from The Atlantic Monthly Magazine - 7/2008 - Author - Nicholas Carr

NOTES

NOTES

NOTES

NOTES

NOTES

NOTES

NOTES